Israa Sabri AlFurati

Israa Robotic System

Israa Sabri AlFurati

Israa Robotic System

Design and Implement a Localization System with Low Cost Sensors

Noor Publishing

Cover image: www.ingimage.com

Publisher:
Noor Publishing
is a trademark of
Dodo Books Indian Ocean Ltd., member of the OmniScriptum S.R.L Publishing group
str. A.Russo 15, of. 61, Chisinau-2068, Republic of Moldova Europe
Printed at: see last page
ISBN: 978-620-3-85873-0

DESIGN AND IMPLEMENTATION OF INDOOR ROBOT POSITIONING SYSTEM USING LED ARRAY AND LDR SENSOR

Abstract

One of the most challenges in the robotic system is robot localization. It is the first step in robot navigation strategies. Its importance has been noted in a wide variety of applications, An efficient localization system must localize the robot precisely. In this paper, the localization procedure and the structure of the proposed robot are explained. Pairs of low-cost sensors are used to build the suggested system. The light-dependent resistor (LDR) sensors are equipped on the robot base and the light-emitting diodes (LED) sensors are prepared on the environment as a grid of various sizes. The location of each LED in the environment is known since they are connected as a grid and each LED in the grid is identified. The localization process is initialized by turning on the LED array using the modified binary search algorithm. At each step of this calculation, the LDR sensors record their locations as for the LEDs light identified. The modified Binary search algorithm is used to control the turning on the lights of the white-LEDs array to decrease the localization time. The robot localization is accomplished by utilizing the guideline of finding the focal point of a circle with three knowing focuses or the new strategy for finding the focal point of a circle with two knowing focuses and the radius of the circle. This proposed localization framework is actualized practically utilizing a 16*16 LED exhibit and a robot with eight LDR sensors. The localization system is acknowledged on PC by gathering the data of the LDR sensors, which are fitted to the robot. The simulation and experimental results demonstrate good performance in the localization progression.

Keywords: indoor Localization, white LEDs array, LDR sensor, robot.

1. Introduction

Nowadays, a multi-mobile robot is used in many fields such as path planning and obstacle avoidance, object tracking, inspection, surveillance, and multi-mobile robot formation. A localization technique is one of the essential requirements for a multi-robot system which is essential to determine the location and direction of the mobile robot built on one of the localization algorithms and earlier information represents by the environment map or known beacons. However, it is difficult to perform any task by any mobile robot without knowing its location in the environment [1].

Many types of sensors are used for indoor robot localization, including laser range finders (LRFs), ultrasonic sensors, vision sensors, the radio frequency identification (RFID) system, visible light communication (VLC) technology, Bluetooth technology, and Wi-Fi. there are positioning and an (IR) transmitter-receiver. Sensors are small devices that are used to measure or sense near-natural quantities, and through some relationships, they are translated into new movements that people can read for display or additional processing. It is also used for different types of measurements, such as distance, temperature sound, pressure, movement, and light level [2]. With the rapid growth of smart technology, sensors that provide accurate distance detection are looked for. It must also be low-cost. Both the Ultrasonic localization techniques [3, 4] and the LRF [5, 6] are highly accurate in operation. However, these systems have a problem locating a mobile robot accurately if the environment has some unfamiliar moving objects, and the LRFs are restricted in an indoor environment with transparent walls. The RFID system can be used to precisely locate the mobile robot if the environment has a uniform and intense distribution of IC tags [7, 8]. However, this technology is relatively expensive and needs to map the tag IDs and compute these RFID tag locations. Vision sensor-based localization technology [9] can determine robot localization without realigning the environment. Negatively, this technique is affected by changes in environmental lighting. Wi-Fi technologies [10, 11] and Bluetooth [12] can be used to support localization because their small devices can be integrated with mobile devices. Indoor localization using Bluetooth Low Energy (BLE) signals is widely used after the emergence of the BLE protocol. Bluetooth beacons have two roles: broadcast by beacons and reception by sensing devices. Therefore, many researchers have implemented positioning methods using beacons [13]. One of the researchers [14] discovered a channel localization polynomial regression model (PRM), channel separation fingerprinting (FP), outlier detection, and extended Kalman filtering (EKF) for indoor localization using a smartphone with a BLE beacon. This localization algorithm uses FP and PRM to estimate the target location. Explains the basic principles of wireless-based indoor localization and focuses on improving results with new Bluetooth Low Energy technology. This system allows volunteers to create and continuously update radio maps. In this system, a BLE transmitter is installed on the building ground in addition to the current WiFi access point [15]. All of these GPS technologies have major differences in price and accuracy complexity. Also, the interference by other signals affects the localization accuracy of both the WiFi and the Bluetooth techniques. The IR LEDs [16–18] and VLC technology [19, 20] are useful for indoor localization since they have stable signals, fast response, and less affected by environmental variation. This technology suffers from interference between signals in the same environment and the effects of electronics.

Accuracy and inexpensiveness are important because the lifespan of light-emitting diode (LED) sensors is prolonged. Light is expected to last many years. Also, these features make LEDs very suitable for use in localization scenarios, because unlike IR and laser, light is harmless to humans and does not overlap with the effects of electronics in the same environment. Furthermore, the LEDs and LDRs sensors are very cheap compared with using other sensors such as IRs, ultrasonic, leader range finder LRF, and wifi sensor. This paper proposes a practical localization system that overcomes all the limitations in the applications of the above systems. Localization is built on a system with two types of sensors. The first sensor (LDR group) is a receiver sensor fixed to the robot base, and the other one (matrix of white LEDs) is a transmitter sensor which distributed consistently in the system. A differential derives mobile robot is built to be used in testing the localization process. Robots can be identified using at least three of the total recipient (LDR) information using a new suggested equation to estimate the robot position. The paper also proposes a new way to localize the robot using two LDR sensors and the radius of the circle only. The paper is organized as follows: Section 2 details the software design of the proposed algorithm. Section 3 describes the hardware design of the indoor robot localization system. Section 4 describes the simulation results of the conversion algorithm. Section 5 deals with experimental results. Finally, Section 6 displays the conclusion.

2. Software Design of The Robot Localization System

In this paper, a specific calculation for indoor robot localization framework is evaluated utilizing sets of light sensors: LDRs and LEDs. The LDR sensor is fixed around the background side of the robot and acts as a receiver, and the LEDs are represented by a matrix of white LEDs that are transmitted consistently in the environment to act as a light transmitter. The area of each LED in the grid is known by its area in the framework. The robot localization procedure is based on the operation of the LEDs using the calculation of the modified dual search algorithm and the data collected from the LDR sensors is used to assess the robot region. Only three of the detected LDR sensors that are set up on the robot are sufficient to evaluate the robot area by considering a circle goes through these three LDRs where the robot area is at the focal point.

2.1. The White LED Matrix

The localization environment consists of two-dimensional white LED lights that are constantly deployed as shown in Fig. 1. These LEDs are installed in a variety of openings and secured with a simple elegant layer of plastic and a robot with LDR sensors sensing this plastic layer. A variety of LED is filled as part of the transmitter under the translation frame. All the anodes in each row are connected to the one-row line (Row i) as shown in Fig.1, and all the cathodes in each column are connected to one column line (Col. j). Each row in an array of LEDs is connected to one output of the main control unit to turn on and off the LEDs. Each column in this array is connected to one output of a decoder circuit to turn ON and OFF the LEDs by scanning these columns. Some of the outputs on the control unit are used to control the decoding circuit. The principle of scanning is used to turn on and off LED lights, and this depends on the use of the modified binary search algorithm which is used to decrease the time needed to scan the rows and columns in the array of LEDs. The coordinate axis of each LED depends on its row and column locations in the array of LEDs.

2.2. The LDR Sensors

The lower side of the mobile robot is equipped with a group of LDR sensors distributed uniformly in the surrounding of that robot as shown in Fig. 2 These sensors have limit sensing range and represent the receiver part of the light sensors which are used for the robot localization system. The robot console transmits information collected by LDR sensors to the main monitor. The main control unit can estimate the current position of the robot by synchronizing the timing of turning on the LED array and the time to collect the reading of the LDR sensor. The robot localization has occurred when at least three of the LDRs sensors equipped on the robot recognize three LEDs. These three detect LEDs locate at the surrounding of a circle and the robot location is in the center of this circle.

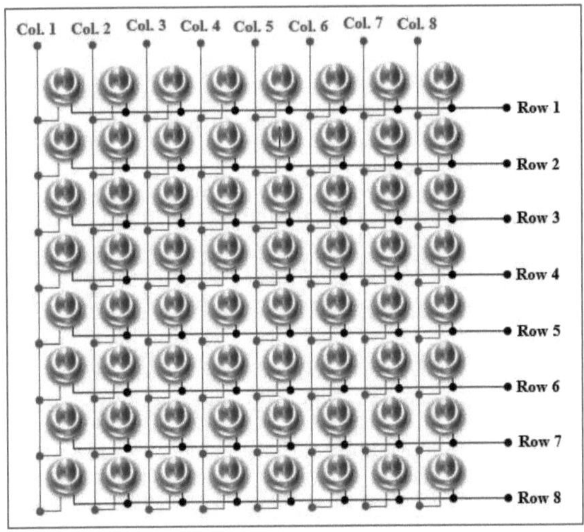

Fig. 1. The 8*8 LED matrix.

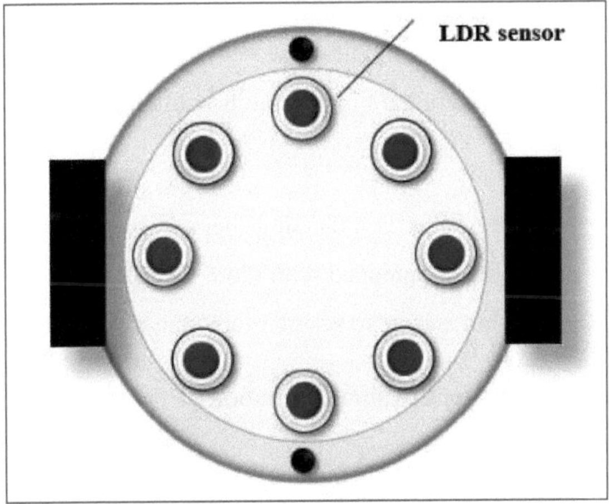

Fig. 2. The lower side of the robot.

2.3. The Modified Binary Search Algorithm

The binary search (logarithmic search) is the most popular search algorithm that finds the location of the target number within a sorted array. This algorithm compares the searching number in the middle of the array. Half below the middle is eliminated when the searching number is not equal to the middle. So, in this case, only the second half is searching by taking the middle number of this half. This process is repeated by splitting the array in half until the number is found. When the searching process ends without finding the searched number, it means that the number is not found. The algorithm needs to remember the start and end of the rest of the array. The complexity of this algorithm depends on the logarithm of the array size [21,22]. The binary search algorithm has been changed to be compatible with the localization system. The system looks for multiple elements, each with a binary value, so the changes are made according to the following differences:

1. The modified algorithm is a two-dimensional search algorithm where the binary search is a one-dimensional algorithm.

2. Each element in the modified search algorithm is represented by a binary value, where the binary search is used for searching a decimal value.

3. Each searching value is indicating the status of the LDR sensor and since each robot is equipped with more than one sensor, the modified search algorithm is used to search more than one value at a time.

Following are the steps of implementation of the modified binary search algorithm:

1. For the columns of LEDs

a. Divide the group of columns into two subgroups.

b. Turn ON each subgroup of LEDs separately.

c. For each subgroup, if any of the LDR sensors detect light, then the status of this subgroup labels to one, while the others label to zero.

d. Repeat the steps a, b and c for each subgroup with label equal to one until reach to subgroup with one column. This process leads to label some LED columns by one and the others of zero depending on the sensitivity of the LDR sensors.

2. For the rows of the LEDs

   a. For each column labeled by one, divide the LEDs of the column into two subgroups.

   b. Turn ON each subgroup of LEDs separately.

   c. For each subgroup, if any of the LDR sensors detect light, then the status of this subgroup labels to one, while the others label to zero.

   d. This process leads to label some LEDs in each column by one and the others of zero depending on the sensitivity of the LDR sensors.

3. The modified binary searching algorithm completes its task of labeling some LEDs location by one. These locations represent the current location of some of the LDR sensors on the robot.

Information collected from LDR sensor readings can be used to estimate robot position. Fig. 3 shows that four of the LDR sensors are localized by labeling four LEDs and this information is used for the robot localization.

2.4. Robot Location Estimation Using Three LDR Sensors

This area discloses the strategy used to appraise the area of the robot utilizing data gathered from three LDR sensors fixed to the robot. The location of these sensors relies upon the location of the LEDs identified by the LDR sensors. The location of each LED is computed according to its row and column location. Thus, easy access to the three LEDs identified is spoken by a virtual circle, the focus of which represents the robot's area. The status of the virtual circle can be calculated using the following equation:

$$Ax^2 + By^2 + Bx + Cy + D = 0 \quad (1)$$

To derive the equations for the center of this circle, three points are assumed: ((x1, y1), (x2, y2) and (x3, y3)) pass through the circumference of the circle as shown in Fig. 4.

By substituting the three points which lie on the circle in the Eq. (1) of the circle, one can derive the Eq.s (2),(3), and (4). For the center point and the radius of this circle [23].

$$x = \frac{(x_1{}^2+y_1{}^2)(y_2-y_3)+(x_2{}^2+y_2{}^2)(y_3-y_1)+(x_3{}^2+y_3{}^2)(y_1-y_2)}{2(x_1(y_2-y_3)-y_1(x_2-x_3)+x_2y_3-x_3y_2)} \quad (2)$$

$$y = \frac{(x_1{}^2+y_1{}^2)(x_3-x_2)+(x_2{}^2+y_2{}^2)(x_1-x_3)+(x_3{}^2+y_3{}^2)(x_2-x_1)}{2(x_1(y_2-y_3)-y_1(x_2-x_3)+x_2y_3-x_3y_2)} \quad (3)$$

$$r = \sqrt{(x-x_1)^2 + (y-y_1)^2} \quad (4)$$

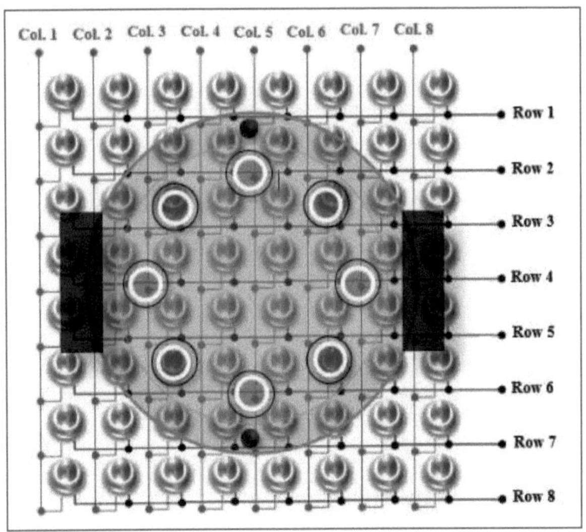

Fig. 3. The robot with four active LDRs sensors localized using four LEDs.

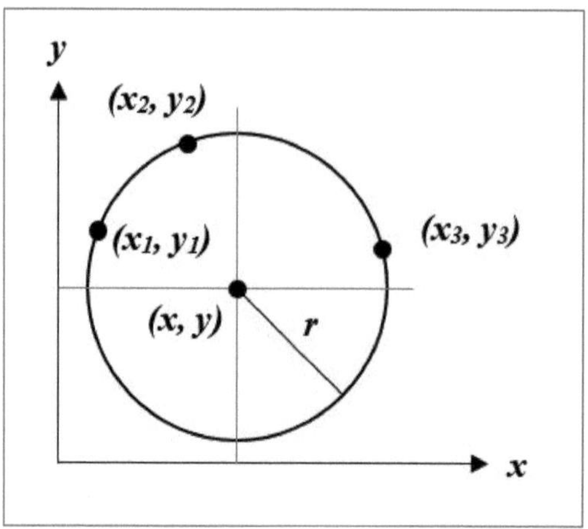

Fig. 4. The virtual circle with three points.

2.5. Robot Localization With Two LDR Sensors Locations and Orientations

This section deals with the estimating of the robot location with two active LDR sensors. The active sensor means the sensor that detects neighbor LED (has the coordinate axis of this LED). All the LDR sensors are located on a circle represented by the surrounding of the robot (Fig. 5) The center of this circle represents the position of the robot. With the radius of this circle known, the center of this circle can be estimated using two active LDR sensors. Two points and radius are not enough to compute the actual location of the circle center because this information leads to an estimate of two circles with two centers as shown in Fig. 6. Since the robot can recognize the name of the active LDR sensors, then this information can be used to distinguish the actual center. The following steps show the procedure used to estimate the robot location using only two active LDR sensors:

Step 1: Assume that N LDR sensors are uniformly distributed around the circumference of the circular robot (LDR1, LDR2…., LDRN) with a radius equal to R. Then the angle α between any two neighbor LDRs are:

$$\alpha = 360/N \qquad (5)$$

For example, when N = 8 (Fig. 5) then the angle α = 45°.

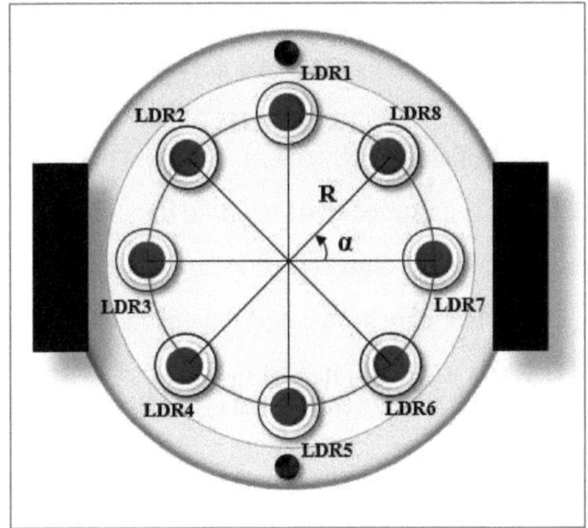

Fig. 5. LDR sensors at the bottom of the robot.

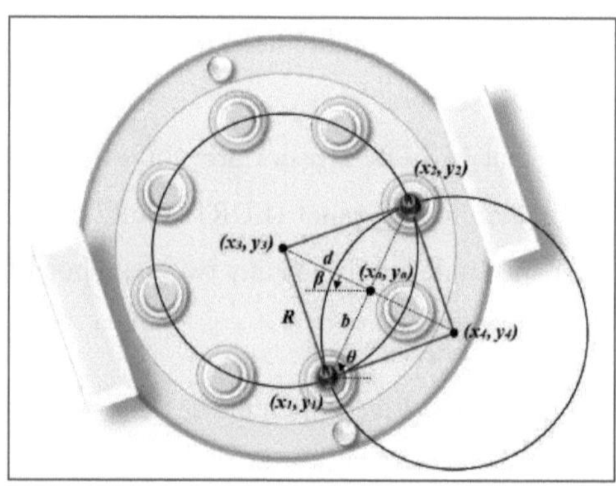

Fig. 6. Drawing two circles using two active LDR sensors and circle radius.

Step 2: When two of the LDR sensors detect two LEDs, then the coordinate axis of these LDRs is assumed to be equal to the coordinate axis of the detected LEDs and these LDRs are represented as active LDRs. As an example, assume two of the LDR sensors are active and the coordinate axis to these sensors are (x1, y1) and (x2, y2) as shown in Fig. 6.

Step 3: Two equal circles with computed center points can be estimated by using only two points and radius where these points pass through the circumference of both the circles [24]. The coordinate axis of the circles center points is computed according to the following procedure:

1. Figure. 6 shows that the two points (x1, y1) and (x2, y2) and the centers of the two circles (x3, y3) and (x4, y4) represent the vertices of a rhombus with each side equal to R. The two diagonals of the rhombus

are perpendicular to each other. The center point of the rhombus is computed according to the following the Eq.s (6) and (7).

$$x_a = x_1 + \frac{(x_2 - x_1)}{2} \quad (6)$$

$$y_a = y_1 + \frac{(y_2 - y_1)}{2} \quad (7)$$

2. The length of line b is computed using Eq.(8) Pythagorean Theorem.

$$b = \sqrt{(x_a - x_1)^2 - (y_a - y_1)^2} \quad (8)$$

Use the Pythagorean theorem to calculate the line length d in Eq.(9).

$$d = \sqrt{R^2 - b^2} \quad (9)$$

The slope of line b is computed by Eq.(10).

$$Slope\ b = tan\,\theta = \frac{(y_2 - y_1)}{(x_2 - x_1)} \quad (10)$$

3. Since the line d is perpendicular to line b, then the slope of this line is

$$Slope\ d = -\frac{1}{Slope\ b} \tag{11}$$

$$Slope\ d = tan\,\beta = -\frac{(x_2-x_1)}{(y_2-y_1)} \tag{12}$$

$$\beta = tan^{-1}\frac{(x_1-x_2)}{(y_2-y_1)} \tag{13}$$

4. The coordinate axis of the two circle centers (x3, y3) and (x4, y4) are computed using Eq.s (14) and(15).

$$x_3 = x_a - d\cos\beta$$
$$y_3 = y_a + d\sin\beta \tag{14}$$

$$x_4 = x_a + d\cos\beta$$
$$y_4 = y_a - d\sin\beta \tag{15}$$

Step 4: One of the two center points represents the actual location of the robot and the other is an incorrect point. The principle is used for arranging the LDR sensors on the robot to help to detect the actual point. As an example, in Fig. 7 when the point (x3, y3) is chosen as the robot location, then by computing the location of the point (x2, y2) using both the center point (x3, y3) and the point (x1, y1). The location of the point (x2, y2) is computed by rotating the point (x1, y1) by 2αo. If the computed point has a coordinated axis equal to the second point (x2, y2) then the chosen point (x3, y3) represents the actual location of the robot as shown in Fig. 7.

Step 5: Fig. 8 represents the incorrect location of the robot ((x1, y1). By repeating the test in step 4 it is found that the computed point has a coordinate axis that differs from the coordinate axis of the second point ((x2, y2).

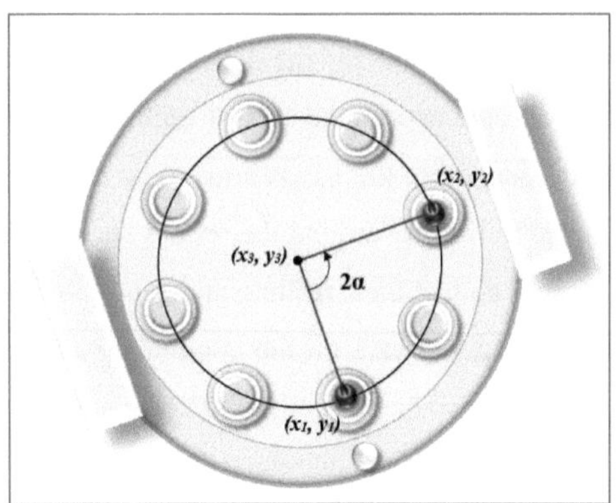

Fig. 7. The actual location of the robot.

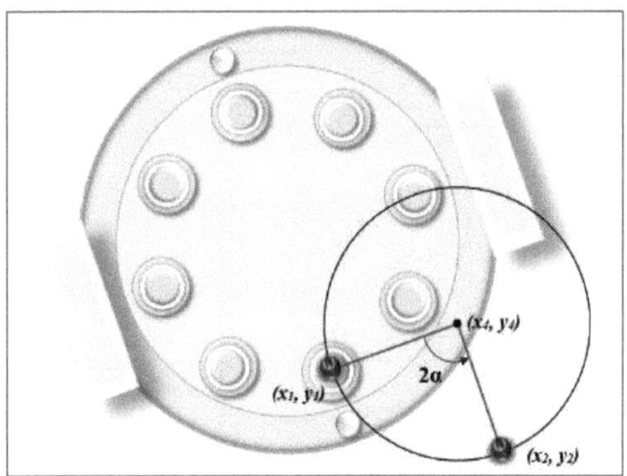

Fig. 8. The incorrect l location of the robot.

3. Hardware Design of The Robot Localization System

This section describes the practical implementation of the robot localization system by scanning a group of lamps in the environment and collecting the reading of the LDR sensors installed on the robot. The diagram of the data collection infrastructure is shown in Fig. 9. There are three main components in this system: the robot localization environment, the robot, and the data logging software. The environment has a matrix of LEDs and the coordinate axis of each LED is determined according to its row and column location in the LEDs array. The robot is equipped with eight LDR sensors distributed uniformly around the robot's surroundings. During the scan process, the LDR sensors collect their reading and send the information to the main computer. The data logging software on the computer estimates the robot location dependent on the reading of two or three of the LDR sensors. The localization system uses an RF transmitter/receiver unit to wirelessly transfer data between the computer on one side and the robot on the other side. The data logging software displays the estimated robot location on the real-time GUI.

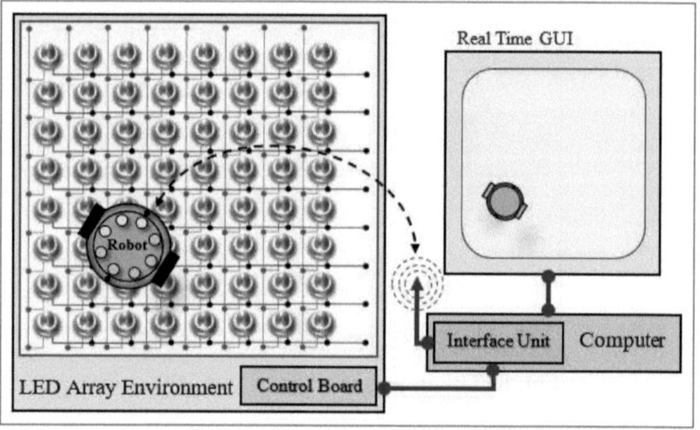

Fig. 9. Scheme of the robot localization system.

3.1. The Robot Designs

The designed robot is a small circular differential drive mobile robot with 10 cm diameter as shown in Fig. 10.a. This robot consists of three parts: The mechanical part which has two servo motors with two caster edges to control the movement of a robot and two 3 v batteries, the main control part which has an Arduino Nano microcontroller to control all the operations on the robot, and LDR sensor part (Fig. 10.b) which has eight LDR sensors to estimate the robot location by sensing the LED array. The main control part receives the orders from the computer to sense the environment when the LED array is scanned. Fig. 11 shows the schematic diagram of the robot's main control part. The robot gives the orders to the LDR sensors to sense the status of the LED arrays and transfer the reading wirelessly (using the NRF24L01 board as shown in Fig.10. a) to the Computer.

(a)

(b)

Fig. 10. Differential drives mobile robots. (a) Robot after adding the wireless communication (NRF24L01) unit. (b) The robot base side

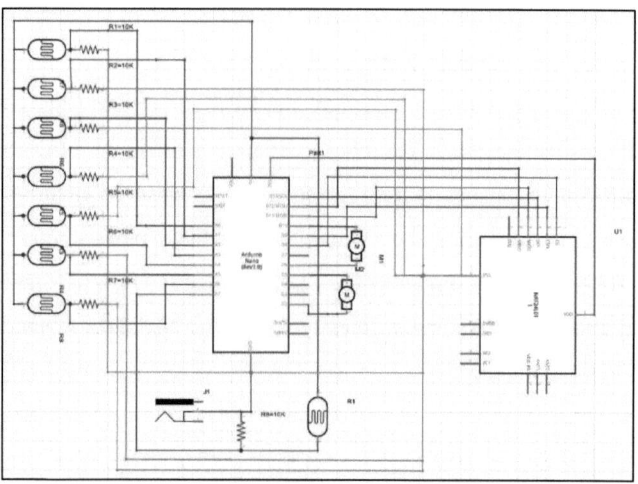

Fig. 11. Schematic diagram of the robot's main control part.

3.2. The Robot Localization Environment

The main component in the localization environment is the LEDs array which is designed as a two-dimensional matrix to fill in as a localization framework as appeared in Fig. 12. These LEDs are fixed in an array of openings and covered by a transparent slim layer of plastic where the robot with LDR sensors moves over this plastic layer. Each row in the array of LEDs is connected to one output of the main control unit to turn ON and OFF the LEDs. Each column in this array is connected to one output of a decoder circuit to turn ON and OFF the LEDs by scan these columns. Some of the outputs on the main screen control are used to control the decoding circuit. The principle of scanning is used to turn on and off LED lights, and this depends on the use of the modified binary search algorithm which is used to decrease the time needed to scan the rows and columns in the array of LEDs. The coordinating axis of each LED depends on its row and column locations in the array of LEDs.

Fig. 12. The robot localization environment.

The main central part of the localization system, which is shown in Fig. 13, is the part that receives the orders from the PC to turning ON and OFF the LEDs using a modified binary search algorithm. This process is done through the NRF24L01 wireless communication unit. The main central part uses the information gathered from the LDR sensors to localize the robot. The LEDs array is divided into four quarters, each one has 8*8 LEDs. Four 74HC574N D F/F are used in this main control unit, each one is used to latch on byte in each quarter of the LEDs array at the same time. To make all the LEDs appear as latch at the same time, a scan principle is used for this purpose. One 74HC138N decoder with eight BD137 NPN transistors is used to investigate the scanning principle. An Arduino UNO microcontroller is used to distribute the information among the four D F/F and also to control the operation of the decoder circuit using the modified binary search algorithm. The collected information from the LDR sensors is transferred wirelessly to the Arduino microcontroller using the NRF Transmitter/Receiver module.

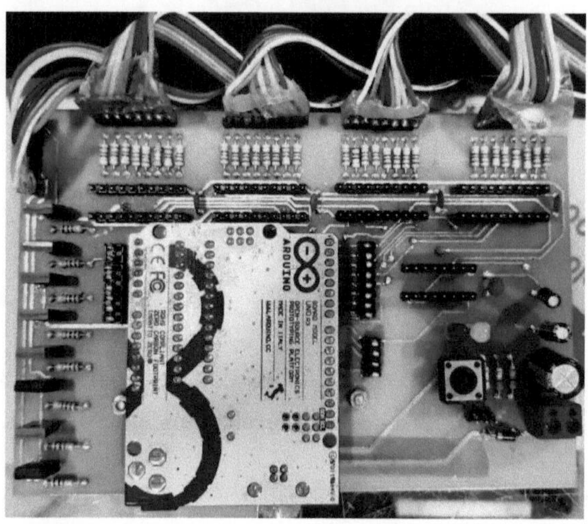

Fig. 13. The main control boards (The electronic circuit).

3.3. Data Logging Software

The computer uses data logging software to collect data from robots and the local environment. The collected data would help in estimating the position of the robot using the modified binary search algorithm. This algorithm determines the position of the robot concerning the LEDs sensed in the local environment. Figure. 14 shows the Graphic User Interface (GUI) for this software. In the LDR section of the GUI, three matrixes are found: the first and the second matrixes have eight rows and two columns of Text Boxes which used to display the actual and the estimated locations of the LDR sensors, and the third one is a one-dimensional matrix used to display the error in estimating the current locations of the LDR sensors. Five Text Boxes are used in GUI to display the current location, the estimated location, and the error in computing the current location of the robot. At the point when push on the RUN Button just the detected LEDs have values in Text Boxes speak to the area of the neighbor LDRs and the undetected LEDs have void qualities in the corresponding TextBoxes. Data logging software speaks to three projects, one is stored in a robot microcontroller, the second one is stored in the microcontroller of the localization framework and the last one is stored in the main control program which stores in the main control unit.

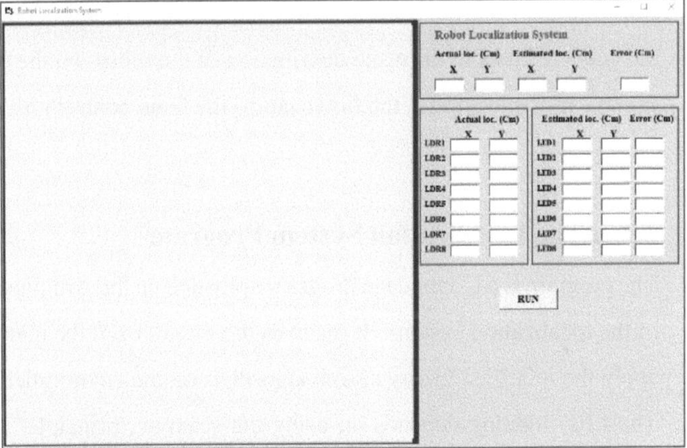

Fig. 14. GUI for the software of the robot localization system.

3.3.1. The Main Control Programs

The main control program is a Visual Basic program that is stored in the main control unit and is used to localize the robots. The main control unit is a microcomputer connected to the microcontroller of the localization framework and a little (NRF24L01) board through the USB links. The (NRF24L01) board has an Arduino microcontroller and (NRF24L01) module to discuss remotely with the robots. The reason for the main control unit is to control the lighting of the LEDs matrix on the localization framework and to get the data collected by LDR sensors on the robot remotely. According to this data, the main control unit displays the area of the robots on it as a GUI. The localization procedure is started by giving the orders to the localization framework to lighting the LEDs according to the modified binary search algorithm. At each progression of the binary search algorithm, the main control unit holds back to get the data collected by the robots which indicate the status of the LDR sensors. After finishing the modified binary search algorithm, the main control unit utilizes the collected data to draw the distribution of the robots on the GUI of this unit. Figure. 15 demonstrates the flowchart to the main control unit.

3.3.2. The Localization System Program

The program is a C-language program uploaded on the Arduino microcontroller of the localization system. It receives the orders from the main control unit to apply the modified binary search algorithm on the environment LEDS matrix. The LED lighting depends on using the scanning principle. The localization system has four D flip flops to mask the information received from the main control unit. The scanning principle is used to distribute the D flip flops information among the columns the LEDs matrix columns by using the decoder logic circuit. This process of LED lighting is repeated until it completes the modified binary search algorithm. Figure. 16 shows the flowchart of the localization system program.

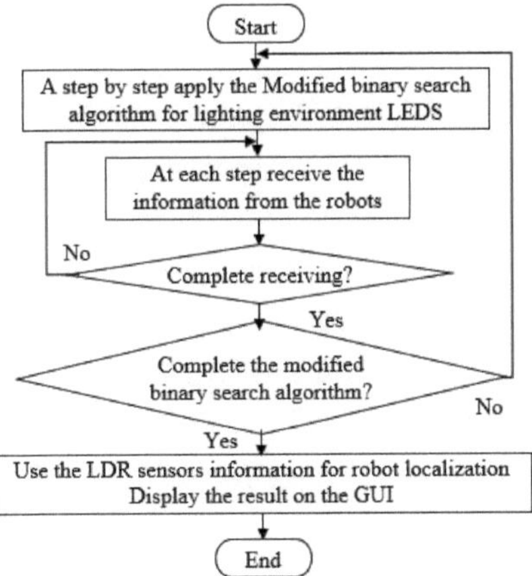

Fig. 15. Flowchart for the main control unit program.

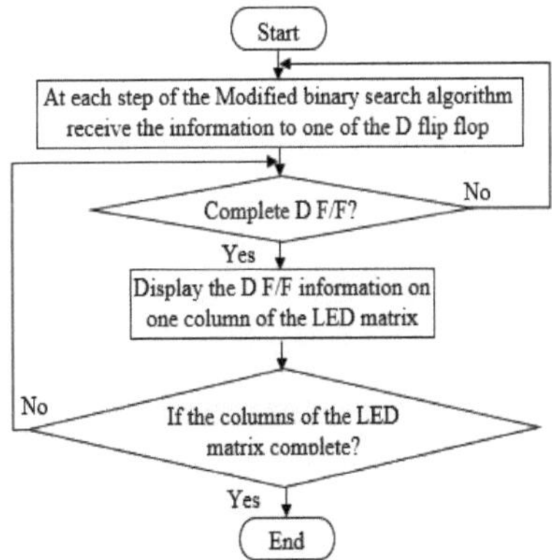

Fig. 16. Flowchart for localization system program.

3.3.3. The Robot Programs

This program is a C-language program store on the Arduino microcontroller of the robot. Every robot has an NRF24L01 module and eight LDR sensors. It gets instructions from the main monitor and starts to read data from the LDR sensor continuously. The collected data is sent to the remote main console. Figure. 17 demonstrates the flowchart of the robot program.

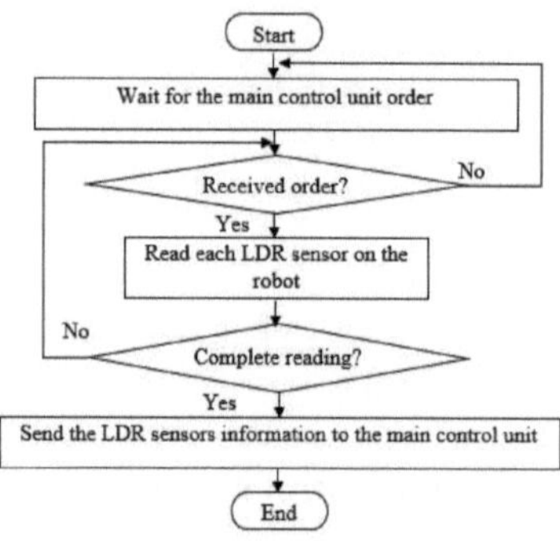

Fig. 17. Flowchart for the robot program.

4. The Simulation Results

This paper simulates another algorithm. The simulations are repeated on a solitary robot with a changed number of LDR sensors (1, 2, 3… 16), different sensing range of these sensors, and different robot areas in an environment with (16×16), (32× 32), (42× 42) and (64× 64) LED matrix. The parameters used in this simulation are:

1) The number of LDR sensors equipped on each robot.

2) The maximum sensing range of the LDR sensor.

3) The number of LEDs in the LED matrix environment.

Figure 18 is used to show the relationship between the full sensor range and sensor range for each LDR sensor in the environment (16 x 16), (32 x 32), (42 x 42), and (64 x 64) LED matrix. The first simulation is performed on a robot with a different number of LDR sensors (1-16) and iterates to a different sensor field of these sensors.

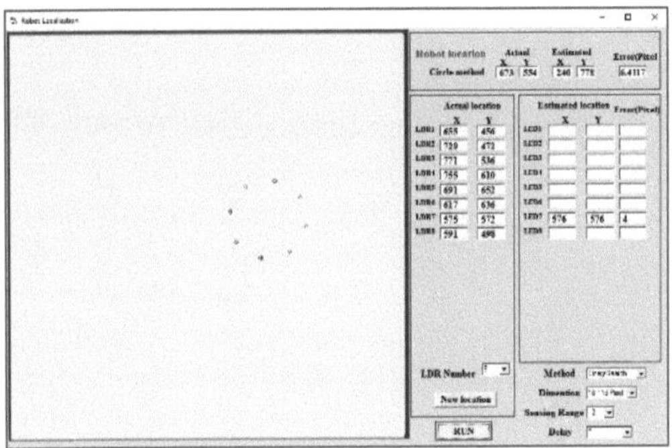

(a)

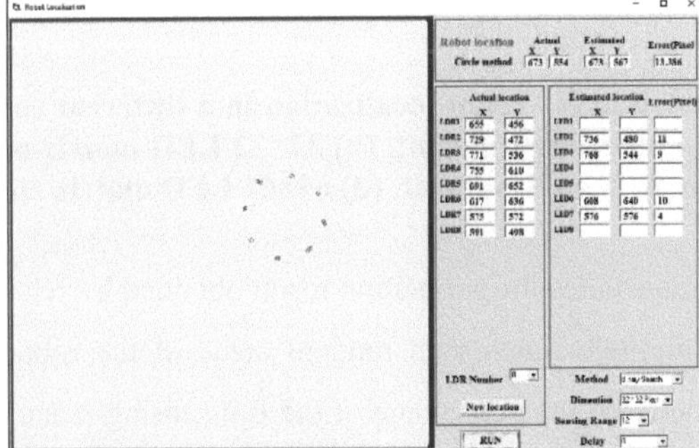

(b)

(c)

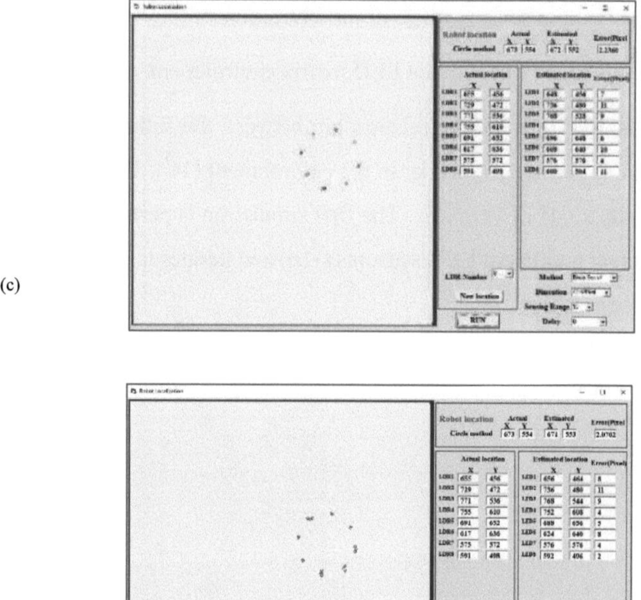

(d)

Fig. 18. Simulation for robot localization in a different environment : (a) 16×16 LED matrix environment. (b) 32×32 LED matrix environment. (c) 42×42 LED matrix environment. (d) 64×64 LED matrix environment.

Figure 19 demonstrates the simulation result obtained by rehashing the above simulations, multiple times with random areas of the robot to acquire the relationship between the percentage of the full sensing range and the number of the LDR sensors for different of the sensing range. The outcome in Fig. 20.b. shows that the robot with eight LDR sensors has the best sensing range to get a full sensing range is 12 Pixels in a 32× 32 LED matrix environment. Likewise, the outcome in Fig. 20.a. demonstrates that the robot with four LDR sensors has the best sensing range to get a full sensing range of 8 Pixels in 64× 64 LED matrix environments.

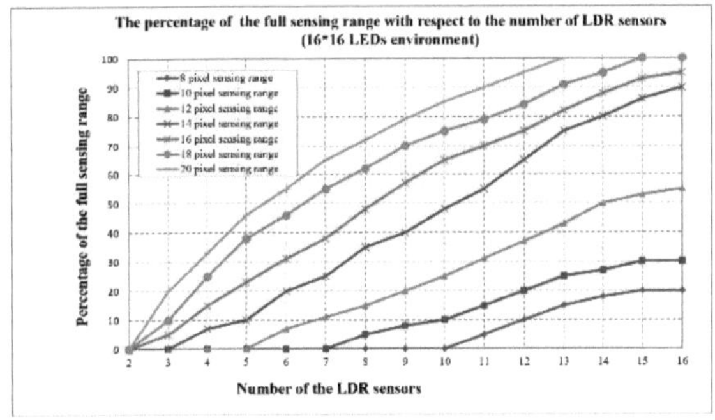

(a)

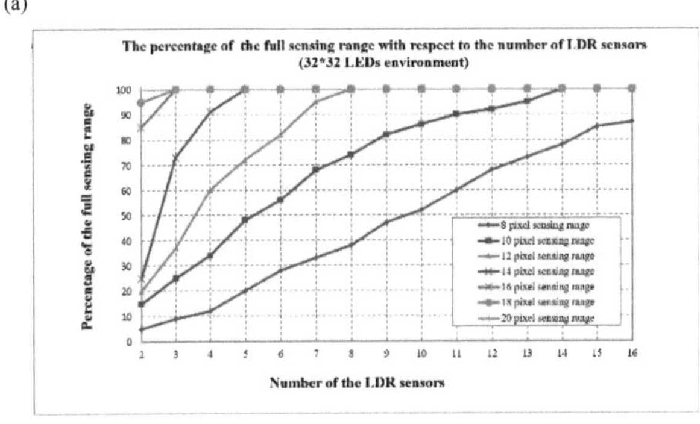

(b)

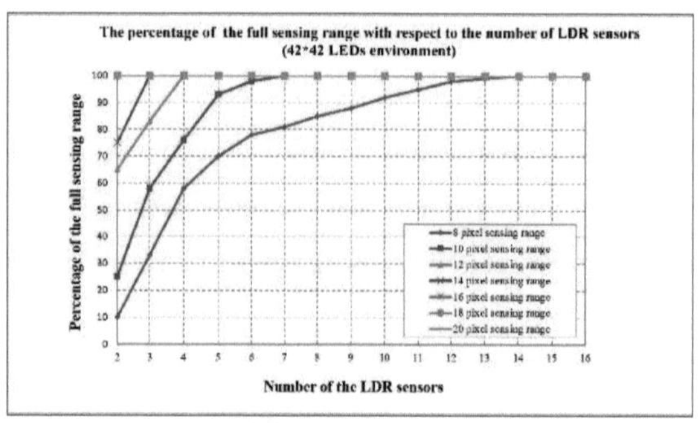

(c)

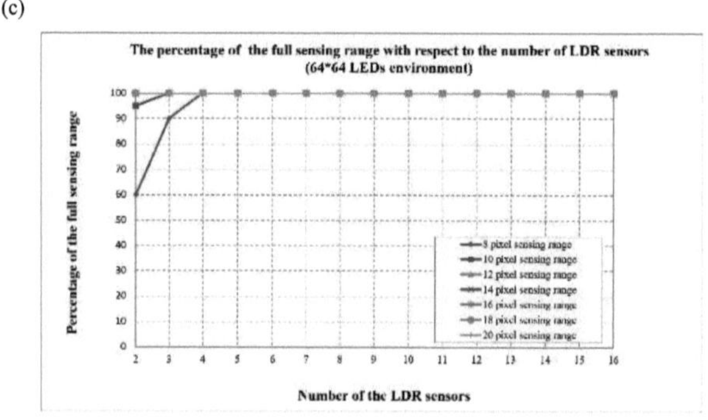

(d)

Fig. 19. Comparison between the ratio of full sensing range and the number of LDR sensors for different sensing ranges. (a) 16×16 LED matrix environment. (b) 32×32 LED matrix environment. (c) 42×42 LED matrix environment. (d) 64×64 LED matrix environment.

The second simulation is used to study the impact of changing the sensing range of the LDR sensors on the average of detecting neighbor LEDs and on the average of the error happen in the robot localization process. Figure 20 is used to demonstrate the relationship between the number of detected neighbor LEDs and the sensing range of LDR sensors (8, 12, 16, and 20 pixels) in the environment which has (64× 64) LED matrix. Figure 21 demonstrates the comparison between the average of the detecting neighbor LEDs and the sensing range of the LDR sensors in (16×16), (32× 32), (42× 42), and (64× 64) LED matrix environment. The results demonstrate that, as the sensing range increment, the average of detecting neighbor LEDs likewise increment and the best case happens when the environment has (64× 64) LED matrix. Fig. 22 demonstrates the comparison between the average of the error in robot localization and the sensing range of the LDR sensors in (32× 32), (42× 42), and (64× 64) LED matrix environments. The results show that as the sensing range decrease, the average error likewise decreases and the best case happens when the environment has (64× 64) LED matrix.

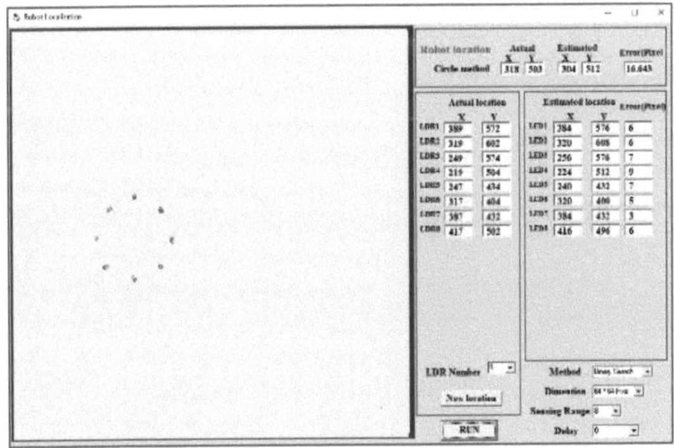

(a)

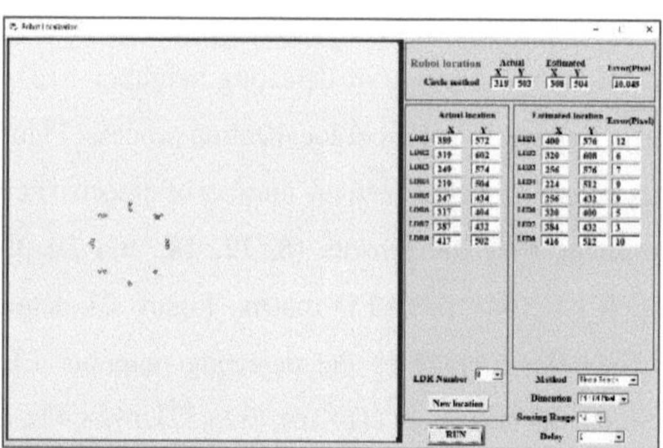

(b)

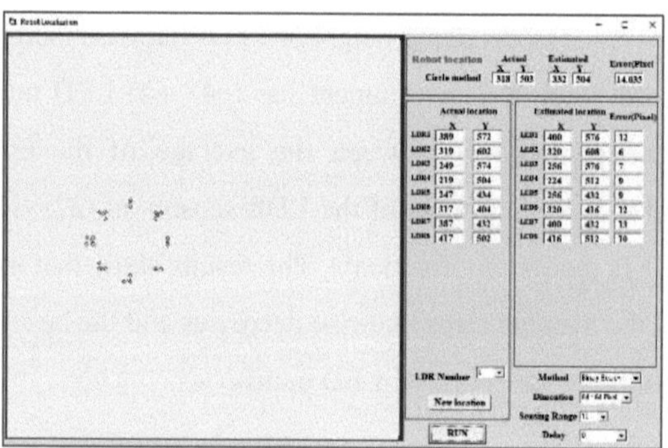

(c)

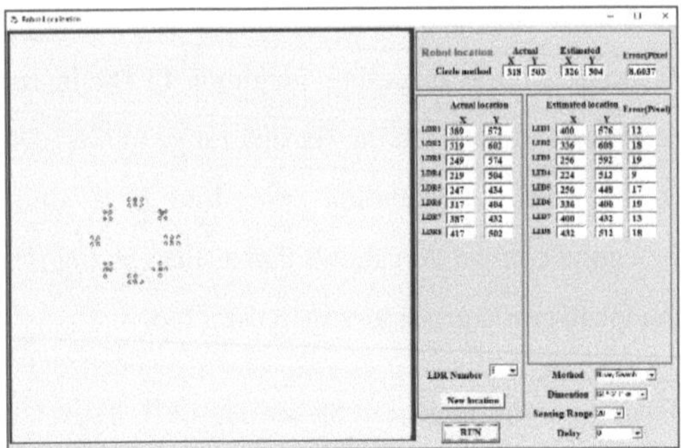

(d)

Fig. 20. The simulation shows the number of detected neighbors LEDs in a 64×64 LED matrix environment. . (a) The sensing range of the LDR sensor is 8 pixels. (b) The sensing range of the LDR sensor is 12 pixels. (c) The sensing range of the LDR sensor is 16 pixels. (d) The sensing range of the LDR sensor is 20 pixels.

5.The Experimental Results

In this area, several experiments for robot localization are executed in (16×16) LED matrix environment (Fig.23) where the distance between every two LEDs is 2 Cm. The LED lighting is controlled by the Arduino microcontroller and the four D flip flops which get the orders from the microcomputer (the main control unit). The lighting procedure is achieved according to the modified binary search algorithm. The experiments in this section are used to study the performance of the localization framework when changing the sensing range of the LDR sensors which prepare on the robot. Figure 24 demonstrates the comparison between the sensing range of the LDR sensors and the average of the detecting neighbor LEDs in (16×16) LED matrix environment. This comparison demonstrates that as the sensing range increment additionally the average of the detecting neighbor LEDs increment. Figure 25 demonstrates the comparison between the sensing range of the LDR sensors and the average of the error in the localization procedure in a similar LED matrix environment. This second comparison shows that as the sensing range increases, the average error in the localization procedure increases, too.

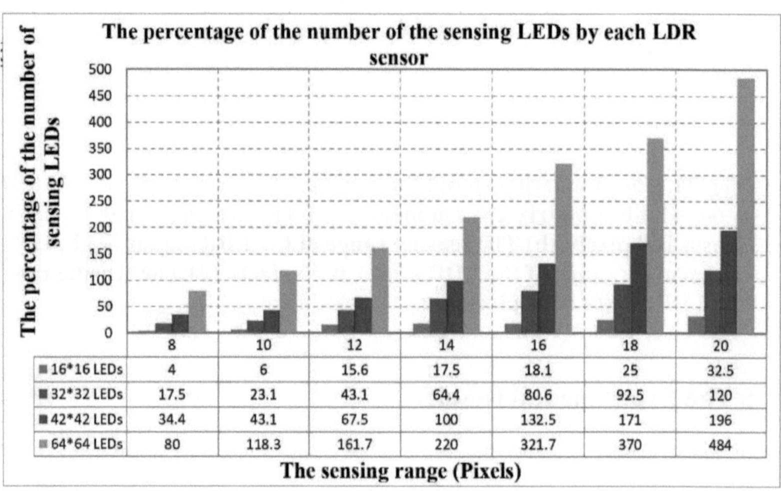

	8	10	12	14	16	18	20
16*16 LEDs	4	6	15.6	17.5	18.1	25	32.5
32*32 LEDs	17.5	23.1	43.1	64.4	80.6	92.5	120
42*42 LEDs	34.4	43.1	67.5	100	132.5	171	196
64*64 LEDs	80	118.3	161.7	220	321.7	370	484

Fig. 21. Comparison between the average of the detecting neighbor LED range in (16×16), (32× 32), (42× 42), and (64× 64) LED matrix environment

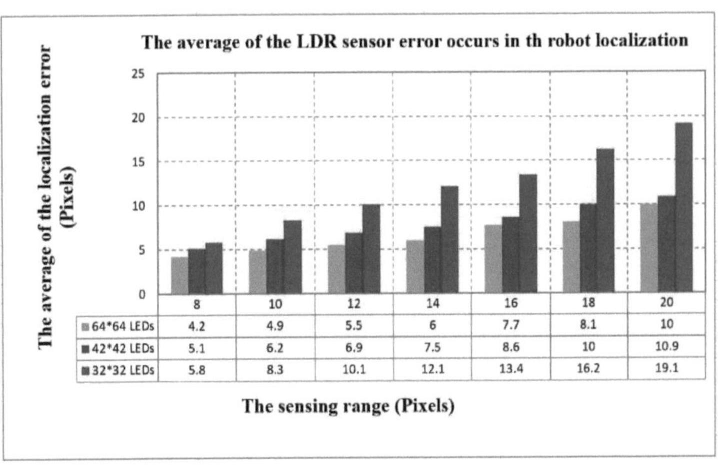

Fig. 22. Comparison between the average of the error in robot localizatio sensing range in (32× 32), (42× 42), and (64× 64) LED matrix environment.

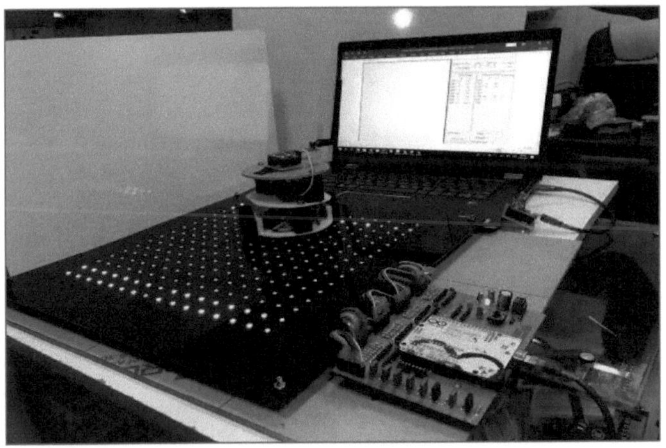

Fig. 23 The practical robot localization environment.

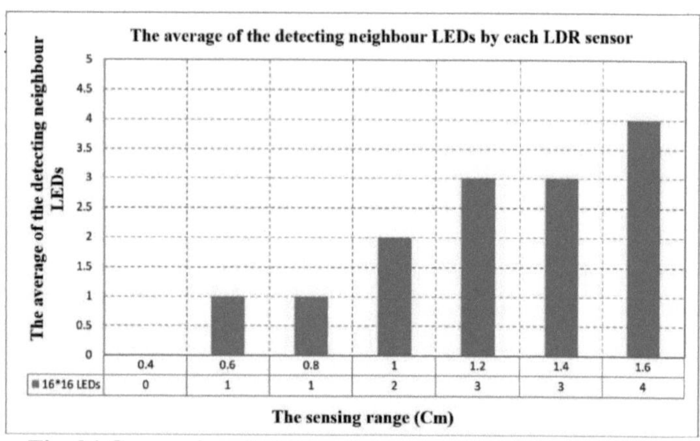

Fig. 24 Comparison between the percentage of the neighbor LEDs and the sensing range in (16×16) LED matrix environment.

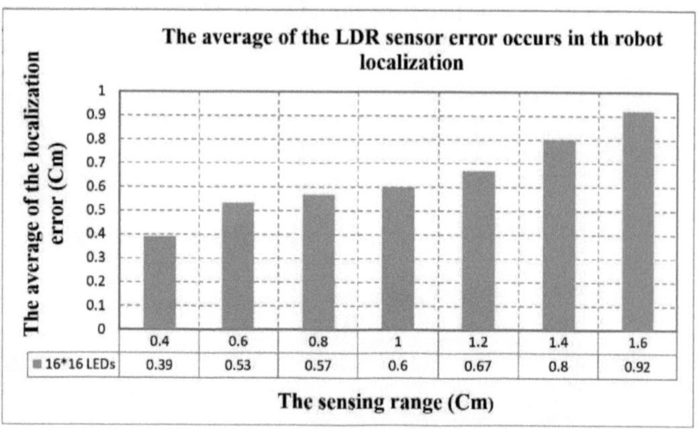

Fig. 25 Comparison between the localization and the sensing range in the average of the error in a robot (16×16) LED matrix environment.

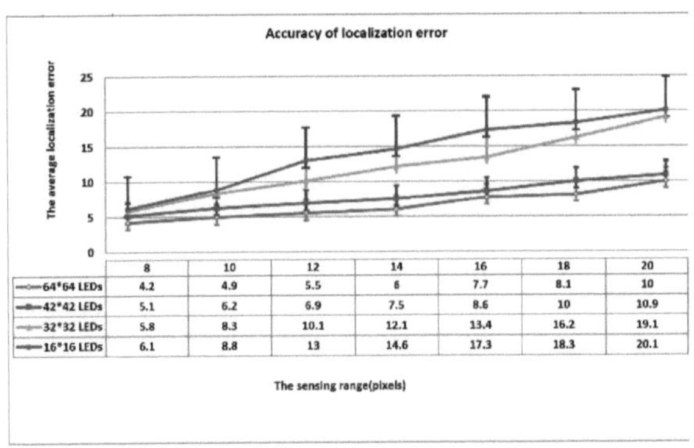

	8	10	12	14	16	18	20
64*64 LEDs	4.2	4.9	5.5	6	7.7	8.1	10
42*42 LEDs	5.1	6.2	6.9	7.5	8.6	10	10.9
32*32 LEDs	5.8	8.3	10.1	12.1	13.4	16.2	19.1
16*16 LEDs	6.1	8.8	13	14.6	17.3	18.3	20.1

Fig.26. Accuracy of robot localization using different sensing range in various environment

6 Conclusion

In this paper, an indoor robot localization framework is presented and executed in both software and hardware. This framework takes care of the issue of localization by utilizing an array of LEDs conveyed consistently in the environment. The localization procedure is actualized and tested on robots with different numbers of LDR sensors (1, 2, 3 … 8 LDR sensors) and different sensing range (8 to 20 Pixels in software and 0.4 to 2 Cm in hardware). In software, the outcomes demonstrate that the best sensing range to accomplish full detection for the LED matrix is 12 Pixels when the robot has eight LDR sensors, and the environment is structured with (32× 32) LED matrix. Likewise, the outcomes demonstrate that the best sensing range to accomplish full detection for the LED matrix is 8 Pixels when the robot has eight LDR sensors and the environment is structured with (64× 64) LED matrix. Additionally, the simulation results demonstrate that the average LDR error is diminished as the sensing range of the LDR sensors decreases. In a common-sense plan, the environment has (16 ×16) LED matrix, and the robot outfitted with eight LDR sensors. The trial results demonstrate that as the sensing range of the LDR sensors is expanding the average of distinguishing neighbor LEDs is increment and the average of the error in the localization procedure expanded.

Conflict of Interest

On behalf of all authors, the corresponding author states that there is no conflict of interest.

Nomenclatures

x	The x-axes for the center point
y	The y-axes for the center point
r	The radius of the circle
N	The number of LDR sensors
xa	The x-axes for the center point of the rhombus
ya	The y-axes for the center point of the rhombus
d	Is the line connected the point (x3,y3) with (xa,ya) (Fig. 6)
R	The radius of the circle (Fig. 6)
$x3$	The x-axes for center point (true location)
$y3$	The y-axes for center point (true location)
$x4$	The x-axes for center point (false location)
$y4$	The y-axes for center point (false location)
$x1$	The x-axes for the first LDR point
$y1$	The y-axes for the first LDR point
$x2$	The x-axes for the second LDR point
$y2$	The y-axes for the second LDR point

Greek Symbols

α	The angle between any two neighbor LDRs, deg.
β	The angle with the axes-axes of the line starts at (xa,ya) (Fig. 6)
θ	The angle result from the intersection of x-axes with the line connected the point (x1,y1) with (x2,y2) (Fig. 6), deg.

Abbreviations

LDR	Light-dependent resistor
LED	Light-emitting diodes
LRFs	Laser range finders
RFID	Radiofrequency identification
VLC	Visible light communication
IR	Infrared sensor
PRM	Polynomial regression model
FP	Fingerprinting
EKF	Extended Kalman filtering

References

1. Rashid, A.T.; Frasca, M.; Ali, A.A.; Rizzo, A.; and Fortuna, L. (2015). Multi-robot localization and orientation estimation using a robotic cluster matching algorithm. *Robotics Auton. Syst.*, 63, 108-121.
2. Soloman, S. (2010). *Sensors and control systems in manufacturing*. New York: McGraw-Hill.
3. Yayan, U.; and Yucel, H. (2015). A low cost ultrasonic based positioning system for the indoor navigation of mobile robots. *Journal of Intelligent & Robotic Systems*, 78(3-4), 541-552.
4. Hasan, O.A.; Rashid, A.T.; Ali, R.S.; and Kosha, J. (2017). Practical performance analysis of low-cost sensors for indoor localization of multi-node systems. *Internet Technologies and Applications (ITA)*, Wrexham, 284-289.
5. Cho, S.H.; and Hong, S. (2010). Map-based indoor robot navigation and localization using a laser range finder. *Proceedings of the 11th International Conference on Control Automation Robotics & Vision*. Singapore, 1559-1564.
6. Wang, Z.; Hwang, Y.S.; Kim, Y.K.; Lee, D.H.; and Lee, J. (2015). Notice of Removal: Mobile robot indoor localization using SURF algorithm based on the LRF sensor. *Proceedings of the 54th annual conference of the society of instrument and control engineers of Japan (SICE)*. Hangzhou, 1122-1125.
7. Mi, J.; and Takahashi, Y. (2015). Performance analysis of mobile robot self-localization based on different configurations of the RFID system. *International conference on advanced intelligent mechatronics (AIM)*. Busan, 1591-1596.

8. Xu, D.; Chen, Y.L.; Lin, C.; Kong, X.; and Wu, X. (2012). Real-time dynamic gesture recognition system based on depth perception for robot navigation. *International Conference on Robotics and Biomimetics (ROBIO).* Guangzhou, 689-694.
9. Lee, J.; Hyun, C.H.; and Park, M. (2013). A vision-based automated guided vehicle system with marker recognition for indoor use. *Sensors,* 13(8), 10052-10073.
10. Jekabsons, G.; Kairish, V.; and Zuravlyov, V. (2011). An analysis of Wi-Fi-based indoor positioning accuracy. *Scientific Journal of Riga Technical University,* 44(1), 131-137.
11. Cline, M.B.; and Pai, D.K. (2003). Post-stabilization for rigid body simulation with contact and constraints. *International Conference on Robotics and Automation* (Cat. No. 03CH37422). Taipei, Taiwan, 3, 3744-3751.
12. Galvan-Tejada, I.; Sandoval, E.I.; and Brena, R. (2012). Wifi Bluetooth based combined positioning algorithm. *Procedia Engineering*, 35, 101-108.
13. Park H.; Noh J.; and Cho S. Three-dimensional positioning system using Bluetooth low-energy beacons. *International Journal of Distributed Sensor Networks*, 12(10), 1-11.
14. Zhuang, Y.; Yang, J.; Li, Y., Qi, L.; and El-Sheimy, N. (2016). Smartphone-based indoor localization with Bluetooth low energy beacons. *Sensors*, 16(5), 596.
15. Kriz, P.; Maly, F.; and Kozel, T. (2016). Improving indoor localization using Bluetooth low energy beacons. *Hindawi*,1-11.

16. Hijikata, S.; Terabayashi, K.; and Umeda, K. (2009). A simple indoor self-localization system using infrared LEDs. *Proceedings of the sixth international conference on networked sensing systems (INSS)*. Pittsburgh, PA, 1-7.

17. Bae, J.; Lee, S.; and Song, J.B. (2008). Use of coded infrared light for mobile robot localization. *Journal of mechanical science and technology*, 22(7), 1279-1286.

18. Mao, L.; Chen, J.; Li, Z.; and Zhang, D. (2013). Relative localization method of multiple microrobots based on simple sensors. *International Journal of Advanced Robotic Systems,* 10(2), 128.

19. Luo, P.; Zhang, M.; Zhang, X.; Cai, G.; Han, D.; and Li, Q. (2013). An indoor visible light communication positioning system using dual-tone multi-frequency technique. *Proceedings of the 2nd International Workshop on Optical Wireless Communications (IWOW)*. Newcastle upon Tyne, 25-29.

20. Qiu, K.; Zhang, F.; and Liu, M. (2015). Visible light communication-based indoor localization using the Gaussian process. *Proceedings of the 28th IEEE/RSJ International Conference on Intelligent Robots and Systems (IROS)*. Hamburg, Germany, 3125-3130.

21. Lieberman, David J.; and Allebach, P. (1997). Efficient model-based halftoning using direct binary search. *In: Proceedings of International Conference on Image Processing. IEEE*, 775-778.

22. NoakO, Robert D (2011). The geometry of generalized binary search. *Transactions on Information Theory, IEEE,* 57.12, 7893-7906.

23. Gross, Dietmar (2017). Statics–Formulas and Problems. *Springer-Verlag GmbH Germany*.

24. Drezner, Z.; Stener, Stefan; Wesolowsky; and George O (2002). On the circle closest to a set of points. *Computers & Operations Research*, 29.6, 637-650.

Content

Printed by Books on Demand GmbH, Norderstedt / Germany